The Power of GPUs in AI and Machine Learning: Driving Modern Advancements

AUTHOR
VISHAL SURATI

Table of Contents

Table of Contents 2
Introduction to GPUs and Their Role in AI and ML 4
Understanding GPUs: Architecture and Functionality 8
The Rise of Deep Learning and Demand for GPUs 12
Key GPU Features That Support AI and ML 18
How GPUs Work in Various AI Models 24
GPUs and Natural Language Processing (NLP) 30
Exploring Cloud GPUs: Access to Supercomputing Power 36
Edge Computing with GPUs for Real-Time AI Applications 44
The Role of GPUs in Advancing Reinforcement Learning 52
Comparing Popular GPU Providers and Their Offerings 62
Recent Advancements in GPU Technology for AI 70
Future of GPUs in AI and Machine Learning 72
Challenges and Limitations of GPUs in AI 76
Conclusion: The Transformative Impact of GPUs on AI and ML 80
Disclaimer 84

Introduction to GPUs and Their Role in AI and ML

Purpose

The purpose of this chapter is to introduce readers to the fundamental role of Graphics Processing Units (GPUs) in artificial intelligence (AI) and machine learning (ML). It will provide background on the evolution of GPUs, explaining how they shifted from gaming applications to being integral to the high-performance computing required in AI.

Key Sections and Concepts

1. **What are GPUs?**

    - Begin with a simple explanation of GPUs. A Graphics Processing Unit is a specialized electronic circuit designed originally to accelerate the processing of images and render 3D graphics in video games.

    - Define some technical elements of a GPU, such as cores (processing units) and parallel processing, which allow GPUs to perform multiple calculations simultaneously.

2. **The Evolution of GPUs: From Gaming to AI**

    - Briefly cover the history of GPUs, starting with their origins in the gaming industry, where they were used for rendering graphics.

    - Explain how researchers and engineers began to realize that the unique architecture of GPUs, with their parallel processing power, could also accelerate complex mathematical calculations in data science.

- Mention how this shift started around the early 2010s with the rise of deep learning and the high demands of training neural networks.

3. **Why Traditional CPUs are Insufficient for AI Workloads**
    - Describe the central processing unit (CPU) and why, although powerful, it is optimized for sequential tasks, handling fewer tasks at high speed.
    - Explain how the structure of CPUs limits their efficiency in data-intensive AI tasks that require processing millions of calculations simultaneously.
    - Use an analogy: CPUs are like "expert problem solvers," focusing on one complex task at a time, whereas GPUs are like "task specialists," handling thousands of simple tasks all at once.

4. **Why AI and ML Require High Computational Power**
    - Define artificial intelligence and machine learning briefly and discuss why they require significant processing power, especially for training large models.
    - Introduce deep learning and explain how it relies on vast amounts of data, making traditional CPUs inadequate for the workload. Highlight the difference in computation time between CPUs and GPUs for deep learning tasks (for example, training a neural network on a CPU could take weeks, while it could take hours on a GPU).

- Include examples such as image recognition models, which need rapid processing to analyze and learn from thousands of images, a process that can only be done efficiently on a GPU.

5. **The Role of GPUs in Today's AI Revolution**
    - Emphasize that GPUs have become foundational to modern AI. Today, they're found in everything from high-performance computers to cloud services, driving AI progress across industries.
    - Briefly introduce the use of GPUs in everyday AI applications, such as voice recognition, autonomous driving, and personalized recommendations in online platforms.
    - Mention how advances in GPU hardware, like the NVIDIA A100 or Tensor Cores, specifically target AI and machine learning workloads, further enhancing the efficiency of these models.

6. **The Future Potential of GPUs in AI**
    - Conclude with a glimpse into the future, suggesting that GPUs will continue to play a key role in scaling AI capabilities, especially as models grow larger and more complex.
    - Mention areas of research and potential challenges, like energy consumption, and touch on how the field is exploring alternative processing methods (e.g., quantum computing, custom AI hardware) to complement GPUs.

Understanding GPUs: Architecture and Functionality

Purpose

This chapter will dive deeper into the technical aspects of GPU architecture and functionality. It will cover how GPUs are built, why they're optimized for parallel processing, and how their architecture differs from that of CPUs, making them ideal for the high-demand tasks in AI and machine learning. This chapter will give readers a foundational understanding of how GPUs work under the hood, setting the stage for understanding how they accelerate complex AI models.

Key Sections and Concepts

1. **Overview of GPU Architecture**

    - Begin with a broad overview of GPU architecture, explaining that a GPU consists of many small cores designed to execute thousands of calculations simultaneously.

    - Describe the main components of a GPU, including cores (or "CUDA cores" for NVIDIA GPUs), memory, and cache. Explain how each of these components contributes to a GPU's ability to handle parallel tasks.

    - Mention that GPUs were originally built to handle high-demand graphics rendering but have become a staple in AI due to this parallelized structure.

2. **Parallel Processing in GPUs vs. Sequential Processing in CPUs**
    - Define parallel processing and explain why it's essential for handling large datasets and complex calculations, like those needed in machine learning.
    - Contrast parallel processing with the sequential processing of CPUs. While CPUs focus on completing one task at a time at high speed, GPUs handle thousands of tasks simultaneously but at slower speeds per task, which is ideal for tasks that can be broken down into smaller parts.
    - Illustrate this with an analogy: CPUs are like fast, single-lane highways, whereas GPUs are like multi-lane highways with many cars moving at once.

3. **The Anatomy of GPU Cores and CUDA**
    - Introduce CUDA (Compute Unified Device Architecture), NVIDIA's platform for parallel computing, which allows developers to take full advantage of GPU power.
    - Explain how CUDA cores operate in parallel to perform calculations, enabling faster matrix operations and data transformations that are essential for AI and ML workloads.
    - Briefly cover the concept of "warps" in CUDA cores (groups of threads executed simultaneously) and how this enables GPUs to manage multiple tasks effectively.

4. **Memory and Bandwidth in GPUs**
    - Describe the high memory bandwidth of GPUs, allowing for quick data movement between cores and memory, which is vital for processing large datasets in ML.
    - Explain how the memory architecture of GPUs is structured to support rapid data access. Contrast this with CPU memory architectures, which are generally optimized for smaller, quicker data access rather than the high-volume transfers needed in AI.
    - Mention that GPU memory bandwidth allows AI applications to manage and manipulate large matrices, essential in training neural networks and image recognition models.

5. **Tensor Cores and Specialized AI Hardware**
    - Introduce tensor cores (found in recent NVIDIA GPUs like the Volta and Ampere architectures) as specialized hardware components designed to perform tensor calculations, a core part of deep learning operations.
    - Describe how tensor cores optimize matrix multiplications and enable faster deep learning training, making them ideal for large neural networks.
    - Note that tensor cores are specifically aimed at AI and ML applications, setting these GPUs apart from standard gaming GPUs and enhancing their performance in AI workloads.

6. **Comparing GPU and CPU Architectures in AI Workloads**
    - Provide a detailed comparison of GPU and CPU architectures, focusing on their use in AI. Highlight how CPUs are general-purpose processors with fewer cores, while GPUs are specialized processors with thousands of smaller cores optimized for AI tasks.
    - Explain why the architecture difference makes GPUs more suitable for tasks like training large neural networks, where high throughput is more critical than low latency.
    - Offer a practical example, such as comparing the time taken to train a neural network on a CPU versus a GPU, to illustrate the efficiency of GPU architecture in AI.

7. **Energy and Efficiency in GPU Processing**
    - Briefly discuss the trade-offs between energy consumption and processing power in GPUs. While GPUs consume more power due to their architecture, they perform tasks faster, leading to higher efficiency per task in data-heavy processes like AI.
    - Highlight ongoing advancements in GPU design aimed at improving energy efficiency, such as the development of AI-specific hardware like tensor cores, and mention other power-efficient GPU options for AI applications.

8. **Future Directions: Innovations in GPU Architecture for AI**

- Wrap up with a glimpse into future GPU architecture developments that aim to meet the growing demands of AI, such as the integration of even more specialized cores or new memory technologies.
- Mention other emerging processing units designed for AI, like TPUs (Tensor Processing Units) developed by Google, and the potential for further GPU optimizations in areas like quantum computing and neuromorphic processors.

The Rise of Deep Learning and Demand for GPUs

Purpose

Here, we focuses on how the rise of deep learning has fueled the demand for GPUs. This chapter will explain what deep learning is, why it requires significant computational power, and how GPUs became central to training complex neural networks that form the backbone of AI applications like image recognition, natural language processing, and recommendation systems. It will emphasize the synergy between deep learning algorithms and GPU processing capabilities.

Key Sections and Concepts

1. **Introduction to Deep Learning**

    - Begin by defining deep learning as a subfield of machine learning that involves neural networks with many layers, allowing systems to model complex patterns in data.

    - Describe the foundational concepts of deep learning, including the structure of neural networks, neurons, and how information flows through these networks.

    - Briefly introduce different types of deep learning models, such as Convolutional Neural Networks (CNNs) for image recognition and Recurrent Neural Networks (RNNs) for sequence processing, setting up later discussions on GPU applications in specific AI models.

2. **Why Deep Learning Requires High Computational Power**

    o   Explain that deep learning involves processing massive amounts of data and training models on this data, which can involve millions or even billions of parameters.

    o   Describe the concept of backpropagation in neural networks, where errors are used to adjust weights across many layers, requiring numerous matrix multiplications—a task ideally suited for GPUs.

    o   Provide a real-world example, such as training an image recognition model on a large dataset like ImageNet, which would take much longer on a CPU compared to a GPU due to the computational complexity.

3. **How GPUs Accelerate Deep Learning Training**

    o   Explain that GPUs excel in performing the large number of matrix calculations and tensor operations required by deep learning. Each layer in a neural network can involve matrix operations that are naturally suited to parallel processing.

    o   Highlight how GPU parallel processing allows for simultaneous calculations across thousands of cores, significantly reducing the training time for deep learning models.

    o   Use a practical example comparing the training time of a neural network on a CPU versus a GPU, showing how GPUs can speed up training from weeks to days or even hours for large-scale models.

4. **The Role of GPUs in Different Deep Learning Models**

    o   Describe how specific types of deep learning models leverage GPU power:

        - **Convolutional Neural Networks (CNNs)**: Used extensively for image processing and computer vision. Explain that CNNs benefit from GPUs because they involve operations like convolutions that can be parallelized effectively.

        - **Recurrent Neural Networks (RNNs) and Long Short-Term Memory (LSTM) Networks**: Used for sequence-based tasks, such as language translation and speech recognition. Discuss how GPUs speed up the training of RNNs by processing multiple sequences in parallel.

        - **Transformer Models**: Mention transformers, like BERT and GPT, which rely on massive parallel processing capabilities that GPUs provide to handle vast amounts of text data simultaneously.

5. **Case Studies: Real-World Applications of GPUs in Deep Learning**

    o   Provide specific examples of deep learning applications where GPUs are essential:

- **Image and Video Processing**: Examples include facial recognition, medical imaging, and autonomous vehicles. GPUs enable fast processing of high-dimensional image data required in these fields.

- **Natural Language Processing (NLP)**: For applications like sentiment analysis, translation, and conversational AI, GPUs allow rapid training of large NLP models, making real-time applications possible.

- **Recommendation Systems**: Platforms like Netflix and Amazon use GPUs to power recommendation algorithms, processing massive amounts of data to provide personalized recommendations.

6. **The GPU Revolution in Deep Learning Research**

    - Explain how the availability of powerful GPUs, such as NVIDIA's Tesla and V100, has led to breakthroughs in AI research. Many state-of-the-art deep learning models and advancements in AI were only possible due to the computational power GPUs provide.

    - Highlight the role of companies like NVIDIA, which designs GPUs specifically for deep learning, and discuss how NVIDIA's CUDA and Tensor Cores have optimized GPU hardware for AI research.

- Mention the impact on open-source communities and frameworks like TensorFlow and PyTorch, which are optimized to run on GPUs, making deep learning more accessible to researchers and developers.

7. **Challenges and Future Directions in GPU-Driven Deep Learning**
    - Briefly touch on the challenges posed by the increased demand for computational power, such as high energy consumption and the cost of high-performance GPUs.
    - Mention emerging technologies that might complement or expand on GPU capabilities, like Google's TPUs (Tensor Processing Units), which are custom-built for AI tasks, and explore how these innovations might shape the future of deep learning.
    - Introduce the idea of distributed GPU training and cloud-based GPU resources (e.g., from AWS, Google Cloud) as solutions for democratizing access to GPU power, especially for smaller companies and individual researchers.

Key GPU Features That Support AI and ML

Purpose

We will explore the specific features of GPUs that make them powerful tools for AI and machine learning. This chapter focuses on explaining how features like CUDA, tensor cores, and high memory bandwidth enable GPUs to handle complex AI tasks, making them highly efficient for deep learning and other AI applications. Understanding these features will give readers insight into why GPUs are so effective and widely used in the AI field.

Key Sections and Concepts

1. **Introduction to Specialized GPU Features for AI and ML**

    - Start with a brief overview of how GPU manufacturers, particularly NVIDIA, have developed specialized features to optimize GPUs for AI and ML.
    - Mention that these features enhance GPUs' processing efficiency and flexibility in handling tasks that are computationally heavy, like training neural networks and processing large datasets.

2. **CUDA (Compute Unified Device Architecture)**

    - Introduce CUDA, a parallel computing platform and programming model developed by NVIDIA, that enables developers to leverage the power of GPUs for general-purpose processing tasks.
    - Explain that CUDA provides an interface for writing code that runs efficiently on NVIDIA

GPUs, making it easier for AI and ML applications to utilize GPU resources.

- Describe how CUDA enables the parallelization of AI tasks by allowing developers to break down computations into smaller tasks that can be processed simultaneously across the thousands of cores in a GPU.

- Provide a practical example, such as how CUDA speeds up deep learning training in frameworks like TensorFlow and PyTorch by utilizing GPU cores for matrix multiplications and other intensive calculations.

3. **Tensor Cores and Their Role in AI Workloads**

    - Describe tensor cores as specialized hardware units within certain GPUs (starting with the NVIDIA Volta architecture) that are designed to accelerate tensor operations, which are essential for deep learning.

    - Explain that tensor cores are optimized for mixed-precision calculations (combining 16-bit and 32-bit floating-point numbers), which can dramatically increase computational speed while maintaining accuracy in deep learning tasks.

    - Discuss how tensor cores are particularly effective for tasks like matrix multiplications in neural networks, allowing faster training times for models such as Convolutional Neural Networks (CNNs) and Transformer-based models.

- Provide an example, such as comparing training speeds of large language models like GPT with and without tensor cores, to highlight the advantage tensor cores provide in high-demand AI applications.

4. **High Memory Bandwidth and Large Memory Capacity**
    - Explain the significance of high memory bandwidth in GPUs, which enables rapid data movement between cores and memory. This feature is crucial for handling the large datasets typically used in AI and ML.
    - Describe how high memory bandwidth and larger memory capacity allow GPUs to process and store massive datasets without significant delays, making them ideal for deep learning models that require a lot of data to learn effectively.
    - Mention NVIDIA's GDDR6 and HBM (High Bandwidth Memory) technology and how these memory types improve data handling capabilities, especially in high-performance GPUs used in AI research.
    - Provide an example of a high-memory GPU model, like the NVIDIA A100, which is commonly used for training complex AI models, illustrating how this feature reduces training time and increases efficiency.

5. **Support for Mixed Precision Training**
    - Introduce mixed precision training as a technique that uses both 16-bit and 32-bit

floating-point numbers to optimize model training. This reduces memory usage and speeds up processing without significantly impacting model accuracy.

- Explain that GPUs with tensor cores support mixed precision training, enabling AI models to achieve faster training speeds and reduced computational costs.

- Mention that many deep learning frameworks now support mixed precision training, allowing users to take full advantage of this GPU feature to optimize their models.

- Provide an example where mixed precision training is used to speed up training times for large models like ResNet or BERT, showing how this approach allows researchers to experiment more quickly.

6. **Multi-GPU and Distributed GPU Training Support**

    - Discuss the concept of multi-GPU setups, where multiple GPUs work together to train a single AI model, allowing for even faster processing and the ability to handle larger models.

    - Introduce distributed training, where GPU resources are spread across multiple systems (such as in cloud-based solutions) to scale AI models without being limited to a single machine's processing power.

    - Explain how software like NVIDIA's NCCL (NVIDIA Collective Communications Library) enables efficient communication between GPUs in distributed settings, making it easier for large AI projects to scale.

- Mention platforms like Google Cloud and AWS that offer multi-GPU and distributed GPU options, allowing users to run their models on several GPUs simultaneously for faster results.

7. **Flexible AI Framework Integration and Optimizations**

    - Describe how modern GPUs are designed to integrate smoothly with popular AI frameworks, like TensorFlow, PyTorch, and Keras, allowing developers to utilize GPU power without extensive low-level programming.

    - Mention that many of these frameworks are optimized to leverage GPU features like CUDA and tensor cores, making them accessible for developers who may not be GPU experts but want to accelerate their models.

    - Highlight specific libraries, such as cuDNN (CUDA Deep Neural Network library), which optimizes neural network computations on GPUs and makes deep learning workflows more efficient.

8. **The Combined Impact of These Features on AI Development**

    - Conclude the chapter with a summary of how these features work together to make GPUs the best choice for AI and ML tasks.

    - Emphasize that without these specialized features, deep learning advancements would not have been as rapid or accessible, as CPUs alone would not provide the necessary computational power.

- Mention that GPU features like CUDA, tensor cores, and high memory bandwidth will continue to be essential for next-generation AI, enabling faster innovation and wider AI application across industries.

How GPUs Work in Various AI Models

Purpose

Here we focus on how GPUs optimize and accelerate different types of AI models, including Convolutional Neural Networks (CNNs), Recurrent Neural Networks (RNNs), and Generative Adversarial Networks (GANs). Each model type serves unique purposes, from image recognition to language processing to content generation, and GPUs play a key role in handling the computational demands of each. By the end of this chapter, readers will understand why GPUs are invaluable in powering specific AI architectures.

Key Sections and Concepts

1. **Overview of AI Model Types and the Need for GPU Processing**

    - Start with a brief explanation of why different types of AI models are used, emphasizing that each has distinct structures, purposes, and processing demands.

    - Describe the role of GPUs as "accelerators" that make training complex models practical by dramatically reducing processing time and enabling AI models to analyze and learn from data more efficiently.

2. **Convolutional Neural Networks (CNNs) for Image and Video Processing**

    - Introduce CNNs as the primary model type used for image and video recognition tasks, such as facial recognition, object detection, and medical imaging.

- Explain how CNNs use layers of convolutional filters to detect patterns in images, which involves computationally intensive tasks like matrix multiplications and feature mapping that are highly suited to GPUs.
- Describe how GPUs handle these tasks using parallel processing, where each core in the GPU processes a different part of an image simultaneously, enabling faster training and inference times.
- Provide a real-world example, such as how CNNs trained on GPUs allow companies like Tesla to develop autonomous driving systems that can process road images in real time.

3. **Recurrent Neural Networks (RNNs) and Their Variants for Sequential Data**
    - Introduce RNNs, which are used for processing sequential data like language, speech, and time-series data, due to their ability to retain information from previous steps (known as "memory").
    - Explain how RNNs require repeated operations across each data sequence, making them computationally intensive and highly suited to GPUs, which can process multiple sequences simultaneously.
    - Discuss Long Short-Term Memory (LSTM) networks and Gated Recurrent Units (GRUs), which are advanced types of RNNs designed to manage long-term dependencies in sequential data.

- Provide an example of how RNNs and LSTMs powered by GPUs are used in real-time applications, like machine translation (e.g., Google Translate) or chatbots that generate responses based on conversational context.

4. **Transformers and Their Demand for Massive GPU Power in NLP**
    - Describe transformer models, which have become the dominant architecture for natural language processing (NLP) tasks, including BERT, GPT, and T5.
    - Explain that transformers work by processing all elements of a sequence simultaneously rather than sequentially, which requires vast computational resources but enables faster training.
    - Highlight how transformers rely on GPUs to handle attention mechanisms and large-scale parallelism, as they often involve billions of parameters and require high memory capacity.
    - Provide an example of a transformer model like OpenAI's GPT-3, which was trained on thousands of GPUs to achieve state-of-the-art performance in language tasks, from text generation to question answering.

5. **Generative Adversarial Networks (GANs) and Content Generation**
    - Introduce GANs as a type of AI model used for generating new data, such as images, videos, and even realistic human faces, by having two

networks (a generator and a discriminator) that compete to produce better outputs.

- Describe the adversarial training process, where the generator creates fake data, and the discriminator evaluates it, requiring continuous, iterative computations that are computationally demanding and well-suited to parallel processing on GPUs.
- Mention specific applications of GANs, such as generating realistic art, creating virtual environments, or enhancing low-resolution images in real time, all of which require significant GPU power to function efficiently.
- Provide an example of a GAN application, like NVIDIA's GauGAN, which uses GPUs to convert rough sketches into photorealistic images, showcasing the creative potential of GANs in media and entertainment.

6. **Reinforcement Learning (RL) and the Importance of GPUs in Training Agents**

    - Introduce reinforcement learning, where models (known as agents) learn by interacting with an environment to achieve goals through trial and error.
    - Explain that RL training requires running multiple simulations and updating the agent's decision-making policy based on the outcomes, which demands high computational power that GPUs provide through parallel simulation processing.

- Provide a real-world example, such as the use of GPUs in AlphaGo, where reinforcement learning powered by GPUs enabled the AI to play millions of simulated games against itself to master the game of Go.
- Highlight other applications like robotics, where RL-powered agents use GPUs to train in virtual environments before being deployed in the real world.

7. **Case Study: How GPUs Accelerate AI Models in Real-Time Applications**
    - Compile a brief case study that demonstrates how a company or research institution uses GPUs to accelerate various AI models, highlighting the practical impact of GPU power across applications.
    - Example: Autonomous vehicles, which rely on CNNs for real-time image recognition, RNNs for predicting traffic patterns, and RL for decision-making in complex scenarios. Describe how GPUs process this data in real time, making split-second decisions possible.

8. **Challenges in Scaling AI Models and the Future of GPUs in AI**
    - Conclude with a discussion on the challenges of scaling AI models, noting that as models grow in complexity (e.g., with more layers or parameters), they demand even more GPU resources.

- Mention how companies are addressing these challenges, including the use of cloud-based GPU clusters, multi-GPU setups, and emerging specialized AI hardware.
- Look ahead to future advancements in GPU technology, like increased core counts, optimized tensor cores, and higher memory bandwidth, which will continue to support the development of more complex AI models.

GPUs and Natural Language Processing (NLP)

Purpose

We focuses on the transformative role of GPUs in advancing Natural Language Processing (NLP), a field of AI dedicated to understanding and generating human language. This chapter will explain how GPUs support the training and execution of large NLP models, enabling breakthroughs in tasks like translation, sentiment analysis, chatbots, and text generation. By the end of this chapter, readers will understand why NLP is computationally intensive and how GPUs make real-time language processing possible in a range of applications.

Key Sections and Concepts

1. **Introduction to Natural Language Processing (NLP)**
   - Define NLP as a branch of AI that enables computers to interpret, analyze, and generate human language.
   - Briefly describe common NLP tasks, such as sentiment analysis, text classification, language translation, named entity recognition, and text summarization.
   - Explain the importance of NLP in making human-computer interactions more natural and the challenges in processing the nuances, ambiguities, and vast vocabulary of human language.

2. **Why NLP is Computationally Demanding**

- Explain that NLP tasks often involve processing large amounts of text data, requiring extensive computational resources to analyze patterns, structures, and meanings.
- Describe the need for training models on massive text corpora to understand complex language patterns, requiring high memory capacity and extensive processing power.
- Provide a practical example, like training a language model on an entire Wikipedia dataset, which would require high storage, memory, and processing speed to handle the vast vocabulary and sentence structures.

3. **The Role of GPUs in Training Large NLP Models**
   - Discuss how GPUs accelerate the training of NLP models by processing text data in parallel, significantly reducing training times compared to CPUs.
   - Explain that language models involve complex matrix multiplications and attention mechanisms (like those found in transformer models) that are ideally suited for GPU processing.
   - Mention specific frameworks and libraries, such as TensorFlow and PyTorch, that have GPU-optimized functions to process NLP tasks more efficiently.
   - Provide an example comparing the training time for a large NLP model, like BERT, on a CPU

versus a GPU, illustrating the dramatic speed improvement with GPUs.

4. **Transformer Models and the Rise of Attention Mechanisms**
    - Introduce transformer models as a breakthrough architecture for NLP tasks, which use an attention mechanism to process all words in a sequence simultaneously, rather than sequentially, to capture context more effectively.
    - Describe how the attention mechanism in transformers involves computing relationships between all pairs of words in a sequence, which requires high parallel processing power and memory bandwidth—an ideal fit for GPUs.
    - Provide an overview of popular transformer-based models, such as BERT (Bidirectional Encoder Representations from Transformers), GPT (Generative Pre-trained Transformer), and T5 (Text-To-Text Transfer Transformer).
    - Use a real-world example, like how GPT-3's language generation capabilities are powered by thousands of GPUs working together to process and learn from extensive text data.

5. **Practical Applications of GPUs in NLP**
    - **Language Translation**: Explain how transformer models trained on GPUs power translation tools (e.g., Google Translate) that can interpret text across languages with high accuracy and speed,

thanks to GPUs' ability to handle large-scale parallel processing.

- **Sentiment Analysis**: Describe how sentiment analysis models trained on GPUs allow companies to gauge public opinion by analyzing customer reviews, social media posts, and more, in near real-time.
- **Text Summarization and Content Generation**: Discuss how GPUs enable models like T5 or GPT to generate summaries, write articles, or respond to queries in natural language, which is increasingly used in journalism, customer support, and content creation.
- **Chatbots and Virtual Assistants**: Highlight the role of GPUs in supporting NLP models used in virtual assistants (e.g., Alexa, Siri) and customer service chatbots, where GPUs enable quick response times and accurate language understanding.

6. **Case Study: GPUs in NLP at Scale**
    - Provide a case study of how a company or research institution uses GPUs to handle large-scale NLP tasks:
        - Example: OpenAI's GPT-3, a model with over 175 billion parameters, which required thousands of NVIDIA GPUs for training and enables complex tasks like language generation, question answering, and conversational AI.

- Explain how GPT-3 was trained on a wide variety of text sources, using GPUs to learn language patterns, syntax, and context, making it one of the most advanced language models in AI.

7. **Challenges in Scaling NLP Models and GPU Limitations**
    - Discuss the challenges of scaling NLP models due to their high computational and memory demands. Large models often require extensive GPU resources, which can be costly and energy-intensive.
    - Mention limitations such as the high power consumption of GPU clusters and the cost of scaling models on commercial cloud platforms like AWS and Google Cloud.
    - Introduce recent innovations to mitigate these challenges, such as model distillation (compressing models to require less computational power) and pruning (reducing unnecessary parameters), which can make NLP models more efficient on GPUs.

8. **Future Trends in NLP and GPU Advancements**
    - Conclude with a look at future trends, such as the growing popularity of even larger language models and the continued integration of NLP with real-time applications like augmented reality and IoT devices.
    - Highlight potential GPU advancements, like increased memory capacity and tensor core

optimizations, which will further enhance the efficiency of large NLP models.

- Mention emerging AI hardware, such as TPUs and dedicated NLP accelerators, that could complement or extend the capabilities of GPUs in NLP, particularly for on-device and edge computing applications.

Exploring Cloud GPUs: Access to Supercomputing Power

In recent years, cloud computing has revolutionized access to advanced computational resources, bringing supercomputing power to AI and ML developers without the need for expensive hardware investments. Cloud-based Graphics Processing Units (GPUs) have become especially valuable, as they enable businesses, researchers, and developers to harness high-performance GPU power for AI workloads, even if they don't have on-site infrastructure. This chapter will cover how cloud GPUs work, their benefits for AI and machine learning, the major cloud providers and their offerings, and how cloud GPUs have democratized access to powerful AI resources.

The Basics of Cloud GPUs

Cloud GPUs are GPUs provided as a service by cloud platforms like Amazon Web Services (AWS), Google Cloud Platform (GCP), Microsoft Azure, and others. They allow users to rent GPU resources by the hour, on-demand, or through reserved contracts. Cloud GPUs are hosted in data centers, where multiple GPUs are grouped together into powerful clusters and made accessible to users via the internet. By leveraging these resources, developers can train AI models, conduct research, and deploy machine learning applications without needing physical GPUs.

The concept of cloud-based GPUs has become integral to the growth of AI. GPU-intensive tasks, like deep learning and neural network training, require significant processing power, which can be prohibitively expensive for smaller organizations. By offering flexible access to powerful GPUs, cloud providers help remove these barriers, enabling innovation across industries.

Benefits of Using Cloud GPUs for AI and ML

There are numerous advantages to using cloud GPUs for AI and ML tasks, especially for organizations with varying processing needs and limited resources.

1. **Scalability**: Cloud GPUs provide the flexibility to scale computational resources up or down based on demand. For example, an AI team can rent a small number of GPUs for a pilot project and then scale to hundreds or even thousands of GPUs for full-scale production. This flexibility ensures that resources align with project needs without locking companies into expensive hardware purchases.

2. **Cost Efficiency**: Buying high-performance GPUs is a major financial investment, especially for smaller companies or startups. Cloud GPUs eliminate the need for such capital expenses, allowing companies to rent GPUs only when they're needed. Additionally, the pay-as-you-go pricing models on most platforms allow organizations to optimize costs according to their workloads.

3. **Access to High-End Technology**: Cloud providers regularly update their hardware with the latest GPUs, including models like the NVIDIA A100 and H100, which are specifically designed for AI and ML workloads. This gives users access to cutting-edge technology without needing to upgrade hardware on their own.

4. **Reduced Infrastructure Complexity**: Setting up on-premises GPU clusters requires space, power, cooling, and networking resources, all of which add to the complexity and cost of deploying AI models. Cloud providers handle these logistical challenges, allowing

users to focus on their models and applications rather than on infrastructure management.

5. **Global Accessibility and Collaboration**: With cloud GPUs, remote teams around the world can access the same resources, fostering easier collaboration. This is particularly beneficial for international research teams and distributed AI development projects that rely on shared data and model access.

6. **Support for Hybrid and Multi-Cloud Architectures**: Many organizations are adopting hybrid or multi-cloud strategies, combining on-premises infrastructure with cloud services to optimize performance and costs. Cloud GPUs allow AI teams to experiment and deploy on different platforms, providing more flexibility and reducing the risk of vendor lock-in.

Major Cloud GPU Providers and Their Offerings

Several cloud providers offer GPU services specifically tailored for AI and ML workloads. Here's an overview of the top players in the market:

1. **Amazon Web Services (AWS)**:
     - AWS offers Elastic Compute Cloud (EC2) instances with a variety of GPUs, including NVIDIA V100, A100, and T4 GPUs, which cater to different AI needs, from training large models to handling lower-cost inference tasks.
     - AWS also provides specialized instances, such as the p4d instance, which supports large-scale deep learning with NVIDIA A100 GPUs and offers high throughput and memory.

- In addition to on-demand and reserved instances, AWS offers Spot Instances, which allow users to access unused AWS GPU capacity at a discount, making it a cost-effective option for batch-processing and non-urgent tasks.

2. **Google Cloud Platform (GCP)**:
   - GCP provides a variety of GPU offerings through Compute Engine, including NVIDIA T4, V100, and A100 GPUs. It supports flexible configurations, allowing users to add up to 16 GPUs to a single instance.
   - Google's TPUs (Tensor Processing Units) are another option for AI workloads, particularly suited to deep learning tasks, as they are optimized for TensorFlow applications.
   - GCP's AI Platform offers managed ML services with integrated GPUs, simplifying model training and deployment for users who want an end-to-end solution without needing extensive DevOps management.

3. **Microsoft Azure**:
   - Azure provides a range of GPU virtual machines (VMs), including NV-series (NVIDIA Tesla M60), NC-series (NVIDIA Tesla K80 and V100), and ND-series (NVIDIA Tesla P40), which are suited for visualization, training, and inferencing tasks.
   - Azure Machine Learning service enables users to train models at scale using Azure's GPU resources, and it integrates seamlessly with popular machine learning frameworks.

- For users seeking hybrid cloud deployments, Azure offers Azure Arc, which allows organizations to run Azure services on their own infrastructure, providing a flexible way to manage resources.

4. **Other Providers**:
    - **IBM Cloud**: IBM offers NVIDIA V100 and P100 GPUs on demand for AI applications, and its Watson Machine Learning platform provides an easy way to use these GPUs for model training and deployment.
    - **Oracle Cloud**: Oracle provides bare-metal GPU instances with NVIDIA's A100 GPUs, supporting high-performance ML applications and offering cost-effective options with competitive pricing compared to larger providers.

Use Cases: How Cloud GPUs Are Powering AI Applications

1. **Healthcare**: In medical imaging, cloud GPUs enable faster analysis of high-resolution MRI and CT scans, accelerating diagnosis and allowing hospitals to process data without maintaining complex infrastructure on-site.
2. **Finance**: Financial firms use cloud GPUs for fraud detection, risk analysis, and real-time trading models. Cloud GPUs offer the flexibility to ramp up resources for high-demand periods, such as end-of-quarter processing or during market volatility.
3. **Autonomous Vehicles**: Companies in the autonomous driving sector, such as Tesla and Waymo, use cloud GPUs to train large datasets for image recognition,

object detection, and navigation algorithms. Cloud-based GPU clusters allow them to iterate quickly and test models at scale.

4. **Content Creation and Media**: Cloud GPUs power content generation tools, enabling real-time rendering, video editing, and AI-driven image enhancement. Companies like Adobe use cloud GPUs for features like automatic photo tagging and video editing, making professional tools accessible online.

5. **Natural Language Processing (NLP)**: Cloud GPUs enable fast training of NLP models for applications such as translation, customer support chatbots, and virtual assistants. Language models like OpenAI's GPT and Google's BERT rely on GPU clusters for training, leveraging the scalability of cloud GPUs to process large volumes of text data.

Challenges of Cloud GPUs in AI and ML

While cloud GPUs offer numerous benefits, there are some challenges to consider:

1. **Cost Management**: The pay-as-you-go model, while flexible, can become costly if usage isn't carefully monitored. Running GPUs continuously can result in high bills, making budgeting and monitoring essential for sustained cloud use.

2. **Data Security and Compliance**: For industries with strict data regulations, using cloud GPUs may introduce security and compliance challenges, as data must be transferred to the cloud. Hybrid cloud solutions can address these issues by allowing sensitive data to remain on-premises.

3. **Network Latency**: Running AI models on remote GPUs introduces latency, which can be a challenge for real-time applications that require immediate processing, such as gaming or streaming services.

4. **Dependency on Internet Access**: Cloud GPUs require stable, high-speed internet access. In regions with limited connectivity, access to cloud GPU resources may be restricted, impacting the reliability of cloud-based AI solutions.

The Future of Cloud GPUs in AI and ML

As AI models grow larger and require more computing power, the role of cloud GPUs will continue to expand. Cloud providers are expected to introduce more specialized hardware, such as AI accelerators and improved tensor core GPUs, specifically for advanced deep learning and real-time AI applications. Additionally, as multi-cloud and hybrid strategies become more popular, businesses will be able to integrate cloud GPUs seamlessly with their own infrastructure, optimizing performance and cost.

Cloud GPUs are democratizing AI, allowing organizations of all sizes to access high-performance resources without major hardware investments. With continued advancements in cloud GPU offerings, the future holds even greater opportunities for innovation, making it easier for teams around the world to leverage AI and ML for impactful solutions.

Edge Computing with GPUs for Real-Time AI Applications

Edge computing has become a critical component in modern AI deployments, especially for applications requiring real-time processing, low latency, and localized data handling. This chapter explores how GPUs are transforming edge computing, enabling advanced AI capabilities directly at the data source—whether it's in autonomous vehicles, smart cameras, industrial sensors, or IoT devices. We'll dive into how GPU-powered edge computing works, its applications, benefits, and challenges, along with specific examples that illustrate the transformative impact of bringing AI closer to the edge.

Understanding Edge Computing and Its Role in AI

Edge computing involves processing data closer to where it is generated "at the edge" of a network—rather than sending it to a centralized cloud or data center. This approach reduces latency, minimizes bandwidth requirements, and can improve privacy by keeping sensitive data local. For AI applications, which often rely on rapid decision-making and processing, edge computing with GPUs has become essential.

Edge devices with embedded GPUs, like NVIDIA's Jetson series, allow machine learning models to run directly on-site. This setup enables real-time decision-making without the need to send data to the cloud, an especially valuable feature for applications in autonomous vehicles, industrial automation, healthcare, and smart cities.

The Role of GPUs in Enabling Edge AI

GPUs bring several capabilities to edge computing that make them ideal for running AI workloads on local devices:

1. **Parallel Processing Power**: The multi-core architecture of GPUs allows for high-speed parallel processing, which is essential for AI tasks involving image recognition, video analytics, and sensor data analysis.

2. **Efficient Power Usage**: Modern GPUs designed for edge devices, such as NVIDIA's Jetson Nano and Jetson Xavier, are optimized for power efficiency. This feature is crucial for edge environments where devices may run on limited power sources or need to be portable.

3. **Real-Time Inference**: AI models deployed on GPUs at the edge can process data in real time, allowing applications to respond instantly to changes in the environment. This capability is vital for use cases like autonomous driving, where decisions must be made in milliseconds.

4. **Support for AI Frameworks**: Many edge GPUs support popular AI frameworks, including TensorFlow, PyTorch, and ONNX, allowing developers to train and deploy models using familiar tools. NVIDIA's CUDA and TensorRT optimize these frameworks specifically for GPU inference at the edge.

5. **Hardware for Specific AI Tasks**: Advanced edge GPUs now include tensor cores and other specialized hardware for AI, which accelerates tasks like deep learning inference, video processing, and high-resolution image recognition. This hardware allows edge devices to run complex AI models without relying on cloud resources.

Key Applications of GPU-Powered Edge Computing in AI

The use of edge GPUs has enabled a wide array of real-time AI applications. Here are some key examples across different industries:

1. **Autonomous Vehicles**
    - Autonomous vehicles rely on real-time data from multiple sensors, including cameras, LIDAR, and radar, to make split-second driving decisions. GPUs allow these vehicles to process complex sensor data on the fly, identifying objects, predicting movements, and planning routes without relying on a remote server.
    - NVIDIA's Jetson AGX Xavier, for example, is designed to support the high-performance computing demands of autonomous vehicles, enabling advanced driver assistance systems (ADAS) that enhance vehicle safety and functionality.

2. **Smart Surveillance and Security Systems**
    - In smart cities, edge GPUs are used in surveillance cameras to analyze video feeds in real-time, detecting anomalies, recognizing faces, or identifying suspicious activities instantly.
    - GPU-accelerated cameras can perform tasks like license plate recognition, people counting, and behavioral analysis without requiring video data to be sent to the cloud, ensuring quicker response times and reducing data transmission costs.

3. **Healthcare and Medical Imaging**
    - Edge GPUs in healthcare enable real-time processing of medical imaging data. For example, in portable ultrasound machines or MRI scanners, edge computing with GPUs allows doctors to receive instant analysis, assisting in quicker diagnosis.
    - Additionally, wearable health devices with GPU processing can monitor and analyze patient data, such as ECGs or blood oxygen levels, in real-time, alerting healthcare providers to any immediate issues without cloud dependency.

4. **Industrial Automation and Robotics**
    - Industrial environments use edge GPUs to enhance robotics and automation processes, from visual inspection of products to predictive maintenance of machinery.
    - Edge GPUs allow robots to analyze data from sensors and cameras on the factory floor, making decisions autonomously, which can improve manufacturing quality and efficiency. For instance, robotic arms in assembly lines can use real-time computer vision to detect defects, adjusting processes immediately.

5. **Agriculture and Environmental Monitoring**
    - Edge computing with GPUs supports smart farming by analyzing data from soil sensors, weather stations, and drones. AI models can predict crop health, monitor irrigation needs, or detect pest infestations on the spot.

- Drones equipped with edge GPUs can analyze high-resolution images in real-time to assess crop health, detect weeds, or monitor soil conditions, allowing farmers to take immediate action and optimize resource usage.

6. **Retail and Customer Analytics**

    - In retail, edge devices with GPUs analyze in-store camera feeds to track customer movement, predict purchasing behavior, and optimize store layouts.
    - AI-powered digital kiosks and point-of-sale systems use edge GPUs to run recommendation models in real time, providing customers with personalized shopping experiences based on current trends and stock levels.

Benefits of GPU-Powered Edge Computing for AI

1. **Low Latency for Real-Time Applications**: By processing data on-site, edge GPUs eliminate the latency associated with cloud computing. This low-latency capability is essential for applications where delays could lead to safety risks or missed opportunities, like autonomous driving or emergency response.

2. **Reduced Bandwidth and Cost Savings**: Edge computing reduces the need to transfer large volumes of data to the cloud, saving on bandwidth costs and enabling AI in locations with limited connectivity. This is particularly important for video-heavy applications, like surveillance or remote monitoring.

3. **Enhanced Data Privacy and Security**: With edge GPUs, sensitive data can be processed locally, minimizing the risk of data breaches during transmission. For example,

in healthcare and retail, local data processing ensures that personal information remains secure.

4. **Improved Resilience and Reliability**: Edge devices with GPUs can operate independently of cloud infrastructure, making them more resilient in environments where connectivity is unreliable or intermittent. This independence allows for continuous operation, even if cloud access is interrupted.

5. **Energy Efficiency and Cost-Effective Scaling**: Modern edge GPUs are optimized for lower power consumption, making them suitable for battery-powered devices or remote deployments. For organizations, edge computing enables cost-effective scaling, as AI models can be deployed across multiple localized devices without heavy cloud investments.

Challenges of Implementing GPU-Powered Edge Computing

1. **Hardware Constraints and Costs**: Although edge GPUs are optimized for efficiency, their hardware remains less powerful than full-scale data center GPUs. Processing very large models or complex tasks may still be challenging at the edge, and the initial costs of deploying edge GPUs can be high.

2. **Model Optimization Requirements**: AI models need to be optimized for edge devices to fit within their processing and memory constraints. Techniques like model compression, quantization, and pruning are essential but can reduce model accuracy if not handled carefully.

3. **Limited Power and Cooling Options**: Edge devices often operate in environments with limited power supply and

cooling options. GPUs designed for edge applications must be energy-efficient and resilient, as overheating or power failure could disrupt critical operations.

4. **Data Security and Privacy Risks**: Although edge computing enhances data privacy, it still faces security challenges. Data on edge devices may be vulnerable to local attacks, and security protocols must be implemented to prevent unauthorized access.

5. **Scalability and Management Complexity**: Managing multiple edge devices, especially across distributed environments, requires sophisticated software for monitoring, updating, and maintaining models. Edge deployments add complexity to the AI pipeline, requiring robust edge management tools.

Case Study: Edge GPUs in Smart Cities

To illustrate the impact of GPU-powered edge computing, consider the example of smart city projects:

In a smart city, edge devices like traffic cameras, environmental sensors, and public safety systems gather data constantly. Edge GPUs process data locally, analyzing traffic flow, detecting accidents, monitoring pollution levels, and ensuring security. For instance, a traffic camera with an edge GPU can detect congestion or accidents and send real-time alerts to control centers, allowing for faster emergency response without requiring continuous data upload to the cloud.

NVIDIA's Metropolis platform supports these kinds of edge deployments in smart cities. It allows cities to deploy AI models across various locations, with each device analyzing its data independently. This setup reduces cloud dependency, cuts costs, and improves response times, making city operations more efficient and responsive to residents' needs.

The Future of GPU-Powered Edge Computing

As the demand for real-time AI grows, edge computing with GPUs will continue to play a key role. The next generation of edge GPUs promises to deliver even more power efficiency and computational capacity, making them suitable for increasingly sophisticated models. Additionally, advancements in model optimization and edge management software will make it easier to deploy AI at scale on edge devices.

The rise of 5G networks is expected to boost edge computing further, enabling faster data transfer and seamless integration with cloud services. This hybrid approach—combining edge and cloud GPUs—will allow companies to run both real-time, localized AI models and large-scale cloud-based AI workloads, creating a dynamic AI ecosystem that can handle diverse data requirements.

By bringing AI closer to the data source, edge computing with GPUs is unlocking new possibilities across industries, from safer autonomous vehicles to smarter cities. This chapter demonstrates that as edge technology advances, it will continue to reshape how AI operates in the real world, allowing for increasingly responsive, secure, and efficient applications.

The Role of GPUs in Advancing Reinforcement Learning

Reinforcement Learning (RL) represents a distinct branch of machine learning where agents learn by interacting with an environment to achieve specific goals, making it ideal for applications requiring adaptive decision-making, such as robotics, gaming, autonomous vehicles, and finance. The unique demands of RL, including the need for complex simulations, iterative learning, and large-scale computations, have made GPUs essential for its advancement. In this chapter, we'll explore how GPUs power RL, the impact of parallel processing on training agents, the challenges and solutions in scaling RL models with GPUs, and some cutting-edge applications.

Understanding Reinforcement Learning and Why It's Computationally Intensive

Reinforcement Learning is different from other forms of machine learning in that it focuses on training an agent to maximize rewards by taking actions in an environment. Unlike supervised learning, where models learn from labeled data, RL agents learn from the consequences of their actions, adjusting their behavior to improve over time. This iterative process, known as the "trial-and-error" approach, can involve millions of simulations and requires extensive computational resources.

1. **The RL Process and Computational Requirements**
    - In RL, an agent perceives its environment, takes actions, and receives rewards based on the outcomes. Through exploration (trying new actions) and exploitation (choosing known

actions that yield higher rewards), the agent gradually learns a policy to maximize its long-term rewards.

- Training an RL agent requires running numerous simulations, often in parallel, to explore a wide range of scenarios and strategies. In addition to processing power, RL requires high memory capacity to store and recall experiences, as well as fast processing to iterate and improve policies within a reasonable time frame.

2. **Why GPUs are Essential for Reinforcement Learning**

    - GPUs excel at parallel processing, allowing RL algorithms to simulate multiple environments simultaneously, significantly reducing training times.

    - In RL, tasks like updating policies and calculating action values (in deep Q-networks and policy gradient methods) require numerous matrix multiplications and tensor operations. GPUs, with their thousands of cores, can perform these operations more efficiently than CPUs.

    - GPU-optimized frameworks, such as NVIDIA's cuDNN and RAPIDS AI, offer accelerated computations tailored for RL algorithms, further enhancing training speed and scalability.

Key Techniques in Reinforcement Learning and the Role of GPUs

Several RL techniques have been enhanced by GPU capabilities, especially those that involve deep learning components and large-scale parallel processing.

1. **Deep Q-Learning (DQN)**
    - DQN is a popular RL algorithm that combines Q-learning with deep neural networks. Here, the agent learns a Q-function that estimates the expected reward for each action, selecting actions with the highest expected reward.
    - DQNs rely on deep learning models to approximate the Q-function, requiring significant computational power to process vast amounts of data from simulations. GPUs accelerate this process by enabling the parallel training of neural networks and calculating Q-values quickly.
    - An example of DQN in action can be found in video game AI, where the agent rapidly learns optimal strategies by playing games repeatedly, a task made feasible by GPU power.

2. **Policy Gradient Methods and Proximal Policy Optimization (PPO)**
    - Policy gradient methods directly learn a policy that maximizes rewards by adjusting the agent's actions based on probability distributions. This method is often used in continuous action spaces, like robotics.
    - Proximal Policy Optimization (PPO) is a popular RL algorithm that uses a "surrogate objective" to maintain the stability of training, requiring extensive policy updates and environment simulations. The iterative nature of PPO is well-suited to GPUs, as they can perform thousands of simulations and updates in parallel.

- In robotics, PPO with GPU support enables rapid adaptation in real-world scenarios, such as robotic arms adjusting their grip on different objects.

3. **Actor-Critic Methods**

    - Actor-Critic methods combine the strengths of policy gradient and value-based methods by using two networks: the "actor" learns the policy, and the "critic" evaluates the chosen actions.

    - These methods, especially in large environments, require significant computation to keep the actor and critic models updated concurrently. GPUs allow these dual models to be trained simultaneously, enhancing learning speed and stability.

    - Actor-Critic methods are often used in complex multi-agent systems, such as AI-powered financial trading systems, where decisions must be made based on both immediate rewards and long-term strategy.

4. **Multi-Agent Reinforcement Learning (MARL)**

    - MARL involves multiple agents interacting with each other and their environment, making it suitable for cooperative tasks like swarm robotics and competitive tasks like gaming.

    - The complexity of MARL lies in simulating interactions among agents, each with its own policy and learning strategy. GPUs enable the simultaneous training of multiple agents and allow them to learn both independent and coordinated strategies.

- For example, in autonomous drone swarms, MARL with GPU support allows drones to learn collaborative behaviors, like coordinating in search and rescue missions or performing surveillance.

Case Studies: Real-World Applications of GPUs in Reinforcement Learning

1. **Autonomous Vehicles**

    - Reinforcement learning is fundamental to the development of autonomous vehicles, which need to learn safe navigation, obstacle avoidance, and efficient route planning. RL models trained on GPUs simulate real-world driving environments, enabling vehicles to practice decision-making in diverse traffic and weather conditions.
    - NVIDIA's DRIVE platform uses GPUs to run complex RL models that simulate millions of driving scenarios. Through these simulations, the RL agent learns how to react to various situations, which reduces the need for costly real-world testing.

2. **Gaming and Esports**

    - One of the most well-known applications of RL is in AI agents designed to compete in video games, where agents learn advanced strategies that rival human players. OpenAI's Dota 2 agent, OpenAI Five, was trained using PPO on

thousands of GPUs to develop team strategies and individual player skills.
- In another example, DeepMind's AlphaStar agent used multi-agent reinforcement learning with GPU clusters to master the game StarCraft II, learning strategies by playing millions of matches against itself and human players. This required massive computational power to simulate and learn from complex game environments.

3. **Financial Modeling and Trading**
    - In the finance industry, RL agents are used for algorithmic trading, where they analyze financial data and make trading decisions. RL allows these agents to learn from market dynamics, making decisions that maximize profit while managing risk.
    - GPU-powered RL models allow agents to perform real-time analysis of market trends and adapt to changing conditions. By training on historical market data and running simulations, RL models can identify patterns and make predictive decisions in the fast-paced environment of stock trading.

4. **Healthcare and Personalized Treatment**
    - In healthcare, RL models are being explored for personalized treatment recommendations, especially in managing chronic diseases like diabetes, where treatment plans must be frequently adjusted based on a patient's responses.

- GPU-accelerated RL agents can analyze patient data in real-time, optimizing treatment plans based on factors like medication adherence, activity levels, and biometrics. This dynamic approach allows for patient-specific strategies that improve treatment outcomes.

The Challenges of Scaling Reinforcement Learning with GPUs

Despite the advantages of GPUs in RL, several challenges persist, especially as RL applications and model sizes increase.

1. **High Resource Demands and Cost**
   - The need for extensive parallel simulations means that large-scale RL projects require many GPUs, which can be cost-prohibitive for small organizations. Cloud GPU services provide a solution but can lead to high costs over time.
   - Techniques like distributed learning, which spreads workloads across multiple GPUs, and hybrid cloud strategies are often used to manage costs and resources more efficiently.

2. **Model Stability and Convergence**
   - RL models can be unstable and are prone to issues like overfitting or failing to converge, especially in complex environments. GPUs accelerate training but can amplify these issues if the model architecture and training parameters are not optimized.
   - Regularization techniques, model pruning, and hyperparameter tuning are essential for stability in large GPU-based RL models.

Additionally, research into more stable RL algorithms, such as PPO, has been driven by the need to improve convergence rates.

3. **Scalability in Multi-Agent Environments**
    - As the number of agents increases in MARL settings, computational demands grow exponentially, and managing multiple agents' interactions becomes challenging.
    - Techniques such as transfer learning and curriculum learning are being explored to help agents learn from each other's experiences, reducing the amount of unique training data needed for each agent and improving scalability.

The Future of GPUs in Reinforcement Learning

The future of RL with GPUs promises advancements in both performance and applicability. As GPU technology evolves, we expect higher memory bandwidth, more efficient tensor cores, and lower power consumption, all of which will make GPUs even better suited for RL. Additionally, the development of new GPU architectures specifically tailored to RL needs may further improve training speed and stability.

Emerging fields like hybrid cloud-edge RL, where training occurs in the cloud and inference happens at the edge, will allow RL agents to operate in real time while leveraging massive training resources. This approach is already being tested in areas like robotics and smart factories, where real-time learning from dynamic environments is essential.

Quantum computing may eventually complement GPUs in RL by speeding up specific calculations, although practical applications are still years away. In the near term, advancements in GPU-driven reinforcement learning will continue to empower industries to deploy adaptive, intelligent systems that learn and improve over time, enabling solutions that were previously unimaginable.

In summary, GPUs have transformed reinforcement learning by reducing training times, enabling complex simulations, and allowing the development of RL models that learn from high-stakes environments. As GPUs become more powerful and RL algorithms evolve, we'll see even more sophisticated applications that expand the boundaries of AI and automation.

Comparing Popular GPU Providers and Their Offerings

The rapid expansion of AI and machine learning has brought about a competitive landscape among GPU providers, each striving to offer high-performance hardware tailored to the needs of AI researchers, developers, and businesses. While NVIDIA has long held a dominant position, recent years have seen substantial advancements from other major players, including AMD, Intel, and newer AI-specific GPU manufacturers. This chapter will provide a detailed comparison of these leading GPU providers, covering the unique strengths and weaknesses of each, their AI-specific offerings, and the emerging trends in the GPU market.

The Importance of GPU Selection in AI and ML

Choosing the right GPU is crucial for AI and ML projects, as GPUs drive the computational backbone of deep learning and large-scale data processing. Different providers offer unique architectures, optimizations, and specialized features suited to specific types of AI tasks, from training complex neural networks to real-time inferencing. The decision to select one GPU over another can impact model training times, scalability, energy efficiency, and ultimately, cost-effectiveness.

NVIDIA: The Market Leader in AI and ML GPUs

NVIDIA has been a dominant force in the GPU market, with a clear focus on developing hardware and software tailored for AI applications.

1. **NVIDIA's GPU Architecture and Key Features**

- NVIDIA's GPU line is known for its CUDA architecture, a programming platform that enables AI frameworks like TensorFlow and PyTorch to run optimized code on NVIDIA hardware. CUDA allows for efficient parallel processing, essential for the high-dimensional matrix operations common in deep learning.
- NVIDIA GPUs are also equipped with Tensor Cores, which are specialized hardware components introduced in the Volta architecture (with the NVIDIA V100) and further enhanced in the Ampere (A100) and Hopper (H100) architectures. Tensor Cores accelerate deep learning training by performing mixed-precision calculations, balancing speed and precision for large-scale models.
- Memory bandwidth is another strong point for NVIDIA, with high-end models like the A100 and H100 offering substantial memory capacity and high throughput, suitable for processing massive datasets and complex models in natural language processing and computer vision.

2. **Product Line for AI and ML**

    - **Data Center GPUs**: NVIDIA's A100 and H100 are the top choices for enterprise AI, designed to handle extensive training and inference workloads. The A100 is widely used in cloud computing, with support from major providers like AWS, Google Cloud, and Azure.
    - **Jetson Line**: NVIDIA's Jetson series (e.g., Jetson Xavier) is specifically designed for edge computing, with optimized power efficiency for

real-time AI tasks like robotics and autonomous systems.
- **NVIDIA DGX Systems**: DGX systems combine multiple NVIDIA GPUs in a single machine, offering a plug-and-play solution for AI research labs and enterprise teams needing high performance without setting up their own GPU clusters.

3. **Software Ecosystem**
    - NVIDIA's software ecosystem is comprehensive, including tools like cuDNN for deep learning acceleration, TensorRT for model optimization, and RAPIDS for data science workflows. This ecosystem allows developers to leverage the full power of NVIDIA hardware for a wide range of AI and ML tasks.

AMD: An Increasingly Competitive Player in AI and High-Performance Computing

AMD has made significant strides in the GPU market, especially with its Radeon Instinct line, which targets AI, ML, and high-performance computing (HPC).

1. **AMD's GPU Architecture and Key Features**
    - AMD GPUs use the ROCm (Radeon Open Compute) platform, which provides an open-source framework for GPU computing. ROCm is particularly appealing to researchers who prefer open software ecosystems and integrates with major AI frameworks like TensorFlow and PyTorch.

- AMD's GPU architecture, including the RDNA and CDNA families, emphasizes memory bandwidth and computational efficiency, offering competitive alternatives to NVIDIA's hardware for certain AI tasks.
- While AMD lacks a direct equivalent to NVIDIA's Tensor Cores, its Matrix Cores in the CDNA 2 architecture are designed to improve matrix calculations. This feature is useful for deep learning, but it doesn't yet match NVIDIA's specialized hardware for AI.

2. **Product Line for AI and ML**

    - **Radeon Instinct MI Series**: The Radeon Instinct MI50 and MI100 are specifically designed for AI and ML, optimized for data centers and HPC environments. The MI200, based on CDNA 2, is designed to compete directly with NVIDIA's A100, offering high throughput and improved efficiency for training large models.
    - **Versatility for General Computing**: AMD's GPUs are also widely used in gaming and general-purpose computing, making them a versatile option for businesses that may want multi-functional hardware.

3. **Advantages and Limitations**

    - **Advantages**: AMD GPUs are often priced more competitively, making them an attractive choice for startups and smaller AI teams. The open-source ROCm ecosystem also provides flexibility for custom development.

- **Limitations**: AMD's lack of tensor cores or an equivalent hardware accelerator means that its GPUs are generally slower than NVIDIA's for certain types of deep learning, particularly large-scale NLP models.

Intel: Emerging in the AI and ML Market

Intel's entry into the GPU market marks an important step for the company as it seeks to integrate GPUs with its widely used CPUs and other AI accelerators.

1. **Intel's GPU Architecture and Key Features**
    - Intel's Xe architecture is designed for both consumer and data center GPUs. The Xe-HP and Xe-HPC lines are focused on AI and ML, aiming to provide high computational throughput with lower power consumption.
    - Intel's GPUs are designed to integrate seamlessly with Intel CPUs, providing a unified hardware and software ecosystem. This integration allows for effective load balancing between CPUs and GPUs, which can be advantageous in hybrid workloads.
    - Intel also uses the oneAPI platform, which allows developers to write code that runs across different Intel hardware, from CPUs to GPUs to AI accelerators.

2. **Product Line for AI and ML**
    - **Intel Xe-HPC GPUs**: Targeted at data centers, Intel's Xe-HPC line aims to compete with NVIDIA and AMD for large-scale AI tasks. Although still

new to the market, the Xe-HPC line has shown promising performance in certain AI benchmarks, particularly for general-purpose ML tasks.

- **Intel GPUs for Edge AI**: Intel's Movidius Myriad X and other edge-oriented GPUs are optimized for lower-power AI applications, providing an option for IoT and mobile devices where NVIDIA's Jetson would typically be used.

3. **Strengths and Challenges**

    - **Strengths**: Intel's ability to integrate its GPUs with CPUs provides a seamless development experience, especially in data centers already using Intel hardware. The oneAPI platform also provides a unified development framework across Intel's hardware offerings.

    - **Challenges**: Intel's AI-specific GPUs are still relatively new, and they currently lack the maturity and deep learning acceleration features of NVIDIA's offerings. Intel has yet to establish a proven track record in large-scale AI deployments.

Specialized AI Processors: Google's TPUs and Other AI-First Solutions

While not traditional GPUs, specialized AI processors like Google's TPUs (Tensor Processing Units) and other AI-specific accelerators are designed for specific machine learning tasks.

1. **Google's TPUs**

    - Google's TPUs are custom-designed for AI, particularly for tensor operations in deep learning. Available on Google Cloud Platform, TPUs are highly optimized for training and inferencing with TensorFlow.

- TPUs perform well in applications with massive parallel processing requirements, such as large-scale language models and computer vision. They are highly cost-effective for TensorFlow users but may not support other frameworks as flexibly as GPUs.

2. **Other AI-Specific Chips**

    - AI-first companies like Graphcore, Cerebras, and Habana Labs have developed chips specifically for AI workloads, focusing on performance and energy efficiency.

    - These chips, like the Graphcore IPU (Intelligence Processing Unit) and Cerebras CS-1, offer high parallelism and low latency for deep learning tasks, often outperforming traditional GPUs in specific benchmarks. However, their lack of general-purpose functionality limits their use cases compared to GPUs.

Comparative Analysis: NVIDIA vs. AMD vs. Intel vs. AI-Specific Chips

Feature	NVIDIA	AMD	Intel	AI-Specific Chips
AI-Specific Hardware	Tensor Cores, CUDA	Matrix Cores, ROCm	oneAPI, CPU-GPU integration	Tensor Cores, custom AI optimizations
Software Ecosystem	Strong (cuDNN, TensorRT, RAPIDS)	Open-source (ROCm)	oneAPI, Intel Distribution for Python	Varies (TPU for TensorFlow, IPU SDK)
Use Cases	AI/ML, Data Center, Edge	General-Purpose, Data Center	Data Center, Edge AI	Deep Learning (specialized tasks)
Performance	High (best for large-scale ML)	Moderate (cost-effective)	Emerging (limited benchmarks)	High (task-specific efficiency)
Cost	High	Lower	Competitive	Varies (often expensive)

Future Directions in the GPU Market

The GPU market is rapidly evolving, with providers striving to meet the demands of increasingly complex AI models. Key trends include:

1. **Specialized AI Hardware**: NVIDIA, AMD, and Intel are all working on more specialized AI accelerators to handle deep learning tasks with improved efficiency.

2. **Integration with CPUs and Other Accelerators**: Intel's approach to integrating GPUs with CPUs may become more common, especially for workloads requiring a balance of compute power

Recent Advancements in GPU Technology for AI

The field of GPU technology has seen remarkable advancements in recent years, driven by the demands of artificial intelligence and machine learning. These developments have primarily focused on improving processing power, efficiency, and scalability to handle increasingly complex AI models. This chapter provides an overview of the latest GPU innovations, particularly NVIDIA's new A100 and H100 GPUs, AMD's advancements, and emerging trends like energy efficiency and specialized AI processing cores.

NVIDIA's A100 and H100: Redefining AI Performance

NVIDIA's A100 and H100 GPUs, based on the Ampere and Hopper architectures respectively, represent significant leaps in AI performance:

1. **A100 GPU (Ampere Architecture)**: The A100 introduced Tensor Cores capable of mixed-precision calculations, which boost deep learning training and inference by processing lower-precision data alongside higher precision. This enables faster, more energy-efficient training of complex models. Its high memory bandwidth and support for multi-instance GPU (MIG) allow users to split GPU resources across multiple tasks, maximizing productivity and flexibility.

2. **H100 GPU (Hopper Architecture)**: Building on the A100, the H100 GPU is even more optimized for deep learning, featuring advancements like fourth-generation Tensor Cores and support for the FP8 format, which reduces computational demands while maintaining accuracy. The H100 is designed to accelerate transformer models, which are key in NLP, making it ideal for massive models like GPT-3.

AMD's New GPU Releases for AI

AMD has also introduced AI-oriented GPUs, notably the Instinct MI200 series, based on the CDNA 2 architecture. These GPUs focus on high-performance computing and deep learning, offering improved memory bandwidth and better power efficiency. AMD's open-source ROCm platform is compatible with many AI frameworks, allowing developers more flexibility and making AMD a viable alternative to NVIDIA for certain AI workloads.

Key Trends in GPU Technology for AI

1. **Energy Efficiency and Sustainability**: As GPUs grow more powerful, so does their energy consumption. New designs focus on balancing performance with efficiency, integrating power-saving features and supporting hybrid precision to reduce overall energy use.

2. **Specialized AI Hardware**: The trend toward AI-specific hardware, such as tensor cores in NVIDIA GPUs and matrix cores in AMD GPUs, reflects the need for hardware tailored to deep learning. These cores accelerate matrix operations, improving the speed and efficiency of neural network processing.

3. **Scalability and Flexibility**: Recent GPUs support multi-instance capabilities, allowing multiple AI models to run simultaneously on a single GPU. This scalability is essential for cloud providers and organizations with diverse, concurrent AI workloads.

Future of GPUs in AI and Machine Learning

The future of GPUs in AI and machine learning is full of exciting possibilities as the technology continues to advance to meet the demands of increasingly complex models and applications. From quantum computing integrations to new architectures tailored for AI, the future of GPU technology is focused on pushing the boundaries of speed, efficiency, and capability. This chapter explores emerging trends and potential future directions for GPUs, including next-gen AI architectures, innovations in scalability, and the potential role of quantum computing in complementing traditional GPU power.

Emerging Trends in GPU Design for AI

1. **Increased Specialization with AI Cores**: As AI workloads become more specific, future GPUs are likely to include even more specialized AI cores. NVIDIA's Tensor Cores and AMD's Matrix Cores have shown the value of hardware designed to accelerate matrix calculations, so future GPUs may expand these specialized cores to handle tasks like reinforcement learning or advanced natural language processing with increased efficiency and precision.

2. **Hybrid Precision for Efficiency**: GPU manufacturers are increasingly adopting hybrid precision (e.g., using FP8 and FP16 formats) to balance speed and energy efficiency. In the future, GPUs will likely offer even finer control over precision settings, optimizing power usage and enabling high-performance training of ultra-large models while reducing energy costs.

3. **Scalability and Multi-GPU Systems**: As AI projects scale, so will the need for multi-GPU solutions. Future GPUs are expected to integrate better support for multi-GPU

configurations, enabling seamless communication and resource-sharing between GPUs in large clusters. This is crucial for companies handling massive AI workloads or running real-time applications, as multi-GPU systems allow parallel processing across many models simultaneously.

Potential Role of Quantum Computing and GPUs

While quantum computing is still emerging, it holds potential to revolutionize certain aspects of AI. In the future, we may see hybrid systems where quantum processors handle specific tasks, such as optimization problems and high-dimensional data processing, while GPUs handle neural networks and deep learning tasks. This hybrid approach could significantly reduce processing times for certain AI algorithms, creating a powerful synergy between quantum computing and traditional GPUs.

Software and Integration Enhancements

1. **Unified Development Platforms**: The demand for cross-platform AI solutions will likely lead to unified development frameworks, allowing AI developers to seamlessly run models across diverse GPU architectures and cloud environments. NVIDIA's CUDA and Intel's oneAPI are paving the way for easier integration, and we can expect more cross-compatible software solutions that enhance flexibility and accessibility for developers.

2. **Improved Cloud and Edge Integration**: With AI increasingly moving to the cloud and edge, future GPUs will be designed to work harmoniously across cloud infrastructure and edge devices. This will provide greater flexibility for industries needing both real-time edge processing and high-power cloud computing,

allowing them to choose the optimal environment for each task.

Challenges and Limitations of GPUs in AI

While GPUs have become the backbone of AI and machine learning, they are not without their challenges and limitations. As the demand for complex AI applications grows, issues such as high energy consumption, scalability constraints, and cost can hinder widespread adoption and efficiency. In this chapter, we'll explore the primary challenges faced by GPUs in AI applications and some potential solutions that are emerging to address these limitations.

High Energy Consumption and Environmental Impact

1. **Power Demand:** GPUs, especially those used in data centers and large-scale AI projects, consume significant amounts of energy. High-performance GPUs such as NVIDIA's A100 and H100 or AMD's MI200 are power-hungry, contributing to increased operational costs and carbon footprints, particularly when used in large clusters for training deep learning models.

2. **Environmental Concerns:** Energy consumption from GPUs in AI is a growing concern, as the need for energy-intensive data centers can contribute to environmental degradation. This challenge has driven research into making GPUs more power-efficient, but the issue remains a significant limitation, especially for companies aiming to reduce their carbon footprint.

Cost of High-Performance GPUs

1. **Initial Investment and Maintenance Costs:** The cost of high-performance GPUs, along with maintenance

expenses, can be prohibitively expensive for small and medium-sized enterprises. The A100 and similar GPUs can cost tens of thousands of dollars per unit, limiting access to these powerful tools for smaller AI teams and startups.

2. **Cloud GPU Costs:** While cloud providers offer more affordable access to GPU power, running large models continuously on rented cloud GPUs can still become costly over time. The on-demand pricing model can lead to unpredictable expenses, especially for businesses scaling their AI applications.

Scalability Challenges and Latency

1. **Scalability Constraints:** As AI models grow larger, it becomes increasingly challenging to scale GPU resources effectively. Connecting multiple GPUs in parallel requires significant bandwidth and computational resources to manage the workload efficiently, which can lead to bottlenecks.

2. **Latency Issues:** For real-time AI applications, latency can be a limitation. Even though GPUs enable rapid data processing, the communication overhead in multi-GPU systems or cloud environments can introduce latency, affecting applications like autonomous driving or real-time video analytics.

Limited Compatibility and Integration Challenges

1. **Framework Compatibility:** While frameworks like CUDA are optimized for NVIDIA GPUs, there is still limited cross-compatibility between different providers, such as AMD or Intel GPUs. This lack of interoperability

complicates integration across heterogeneous hardware environments, leading to vendor lock-in.

2. **Software Complexity:** Leveraging the full power of GPUs requires technical expertise in programming with CUDA, ROCm, or other GPU-specific libraries. This complexity can limit accessibility for developers and organizations without specialized knowledge in GPU programming.

Potential Solutions and Future Directions

1. **Energy Efficiency Innovations:** Research into hybrid precision and AI-specific cores, such as tensor and matrix cores, is helping to reduce energy usage while maintaining performance. These innovations are critical for making GPUs more sustainable for AI.

2. **Alternative AI Hardware:** New AI-specific chips like Google's TPUs and dedicated AI accelerators from companies like Graphcore and Cerebras are providing alternatives to GPUs, offering tailored solutions for specific AI tasks with improved power efficiency.

3. **Improved Software Ecosystems:** The development of frameworks like oneAPI (from Intel) aims to bridge compatibility issues, enabling AI models to run on multiple types of processors, including CPUs, GPUs, and specialized accelerators, enhancing flexibility for developers.

Conclusion: The Transformative Impact of GPUs on AI and ML

GPUs have fundamentally transformed the landscape of artificial intelligence and machine learning, unlocking the ability to process complex computations at unprecedented speeds. From their origins in graphics rendering to their central role in AI, GPUs have enabled breakthroughs across a wide range of applications, from deep learning to real-time analytics, NLP, and robotics. By leveraging the unique capabilities of GPUs—such as parallel processing, high memory bandwidth, and specialized AI cores—developers and researchers have been able to build models that were once impossible due to computational constraints.

Key Benefits of GPUs in AI and ML

1. **Acceleration of Training and Inference**: One of the greatest contributions of GPUs has been reducing the time required to train and run machine learning models. Neural networks that once took weeks to train on CPUs can now be completed in days or hours, thanks to GPU power. This speed not only accelerates research but also brings AI closer to real-world deployment across industries.

2. **Support for Complex AI Architectures**: GPUs have made it feasible to implement complex model architectures, such as deep neural networks, convolutional neural networks (CNNs), and transformer models. These architectures, critical for advancements in computer vision and natural language processing, require extensive computational resources that only GPUs can efficiently provide.

3. **Enabling Real-Time AI Applications**: In areas where real-time decision-making is essential, such as autonomous vehicles, healthcare diagnostics, and industrial automation, GPUs allow AI models to process data instantly. By reducing latency, GPUs enable applications that require immediate responses, improving safety and efficiency in high-stakes environments.

The Evolving GPU Landscape

The GPU market is constantly evolving, with companies like NVIDIA, AMD, and Intel leading the charge in developing more efficient, powerful, and specialized hardware. Recent advancements like NVIDIA's H100 and AMD's MI200 are pushing the boundaries of what's possible in AI, integrating features such as tensor cores and matrix cores specifically designed for deep learning. Additionally, new players in the AI hardware space, including Google's TPUs and specialized AI processors from companies like Graphcore, are diversifying options and driving further innovation.

Challenges and Future Directions

While GPUs have driven impressive progress, challenges remain. High energy consumption, costs, scalability issues, and integration complexities are ongoing concerns. These limitations prompt the need for more energy-efficient designs, better support for multi-GPU configurations, and expanded compatibility across hardware platforms. Emerging trends, such as hybrid cloud-edge solutions and quantum computing, may offer future avenues to address these challenges and extend the capabilities of GPUs in AI.

Final Thoughts: The Continued Role of GPUs in AI

As AI models grow more complex and applications more demanding, GPUs will continue to play a crucial role in advancing AI technology. Their ongoing development is essential for unlocking new possibilities in AI, from smart cities and personalized medicine to language understanding and climate science. In many ways, the future of AI depends on the evolution of GPU technology, which will continue to be a driving force in transforming industries and improving lives.

In conclusion, GPUs have not only accelerated AI research but have also empowered AI applications to reach into real-world, practical solutions. The partnership between GPUs and AI is poised to deepen, with continued innovations enabling the next generation of intelligent systems, propelling AI to greater heights in the years to come.

Disclaimer

This book is intended for informational purposes only. While every effort has been made to ensure the accuracy of the content, the author and publisher make no representations or warranties regarding the completeness, accuracy, or suitability of the information for any purpose. The content should not be taken as professional or technical advice, and readers are encouraged to consult industry experts or conduct their own research before making decisions based on the material presented. The author and publisher disclaim any liability for any loss or damage resulting from the use of the information in this book.

www.ingramcontent.com/pod-product-compliance
Lightning Source LLC
Chambersburg PA
CBHW070300220526
45465CB00004B/1685